This is a printed compilation for people that enjoy using and working with printed manuals. The information in this compilation is available for free in PDF format directly from Raspberry Pi. This manual is printed in accordance with their CC BY-ND license. This is a third party printing of their documentation by DienstNet LLC. As an extension, all parts of this compilation not covered by the Raspberry Pi license are also provided under the same CC-BY-ND copyright by DienstNet LLC 2023.

Compilation Contents
Raspberry Pi Pico Python SDK
https://datasheets.raspberrypi.com/pico/raspberry-pi-pico-python-sdk.pdf

ISBN 978-1-365-35620-9

This page was intentionally left blank.

Raspberry Pi Pico Python SDK
A MicroPython environment for RP2040 microcontrollers

Colophon

Copyright © 2020-2022 Raspberry Pi Ltd (formerly Raspberry Pi (Trading) Ltd.)

The documentation of the RP2040 microcontroller is licensed under a Creative Commons Attribution-NoDerivatives 4.0 International (CC BY-ND).

build-date: 2022-11-30
build-version: 3a2defe-clean

> **About the SDK**
>
> Throughout the text "the SDK" refers to our Raspberry Pi Pico SDK. More details about the SDK can be found in the Raspberry Pi Pico C/C++ SDK book. Source code included in the documentation is Copyright © 2020-2022 Raspberry Pi Ltd (formerly Raspberry Pi (Trading) Ltd.) and licensed under the 3-Clause BSD license.

Legal disclaimer notice

TECHNICAL AND RELIABILITY DATA FOR RASPBERRY PI PRODUCTS (INCLUDING DATASHEETS) AS MODIFIED FROM TIME TO TIME ("RESOURCES") ARE PROVIDED BY RASPBERRY PI LTD ("RPL") "AS IS" AND ANY EXPRESS OR IMPLIED WARRANTIES, INCLUDING, BUT NOT LIMITED TO, THE IMPLIED WARRANTIES OF MERCHANTABILITY AND FITNESS FOR A PARTICULAR PURPOSE ARE DISCLAIMED. TO THE MAXIMUM EXTENT PERMITTED BY APPLICABLE LAW IN NO EVENT SHALL RPL BE LIABLE FOR ANY DIRECT, INDIRECT, INCIDENTAL, SPECIAL, EXEMPLARY, OR CONSEQUENTIAL DAMAGES (INCLUDING, BUT NOT LIMITED TO, PROCUREMENT OF SUBSTITUTE GOODS OR SERVICES; LOSS OF USE, DATA, OR PROFITS; OR BUSINESS INTERRUPTION) HOWEVER CAUSED AND ON ANY THEORY OF LIABILITY, WHETHER IN CONTRACT, STRICT LIABILITY, OR TORT (INCLUDING NEGLIGENCE OR OTHERWISE) ARISING IN ANY WAY OUT OF THE USE OF THE RESOURCES, EVEN IF ADVISED OF THE POSSIBILITY OF SUCH DAMAGE.

RPL reserves the right to make any enhancements, improvements, corrections or any other modifications to the RESOURCES or any products described in them at any time and without further notice.

The RESOURCES are intended for skilled users with suitable levels of design knowledge. Users are solely responsible for their selection and use of the RESOURCES and any application of the products described in them. User agrees to indemnify and hold RPL harmless against all liabilities, costs, damages or other losses arising out of their use of the RESOURCES.

RPL grants users permission to use the RESOURCES solely in conjunction with the Raspberry Pi products. All other use of the RESOURCES is prohibited. No licence is granted to any other RPL or other third party intellectual property right.

HIGH RISK ACTIVITIES. Raspberry Pi products are not designed, manufactured or intended for use in hazardous environments requiring fail safe performance, such as in the operation of nuclear facilities, aircraft navigation or communication systems, air traffic control, weapons systems or safety-critical applications (including life support systems and other medical devices), in which the failure of the products could lead directly to death, personal injury or severe physical or environmental damage ("High Risk Activities"). RPL specifically disclaims any express or implied warranty of fitness for High Risk Activities and accepts no liability for use or inclusions of Raspberry Pi products in High Risk Activities.

Raspberry Pi products are provided subject to RPL's Standard Terms. RPL's provision of the RESOURCES does not expand or otherwise modify RPL's Standard Terms including but not limited to the disclaimers and warranties expressed in them.

Table of contents

- Colophon 1
  - Legal disclaimer notice 1
- 1. The MicroPython Environment 3
  - 1.1. Getting MicroPython for RP2040 3
  - 1.2. Installing MicroPython on Raspberry Pi Pico 4
  - 1.3. Building MicroPython From Source 4
- 2. Connecting to the MicroPython REPL 7
  - 2.1. Connecting from a Raspberry Pi over USB 7
  - 2.2. Connecting from a Raspberry Pi using UART 8
  - 2.3. Connecting from a Mac 10
  - 2.4. Say "Hello World" 11
  - 2.5. Blink an LED 11
  - 2.6. What next? 11
- 3. The RP2040 Port 12
  - 3.1. Blinking an LED Forever (Timer) 12
  - 3.2. UART 13
  - 3.3. ADC 13
  - 3.4. Interrupts 14
  - 3.5. Multicore Support 15
  - 3.6. I2C 15
  - 3.7. SPI 16
  - 3.8. PWM 17
  - 3.9. PIO Support 17
    - 3.9.1. IRQ 19
    - 3.9.2. WS2812 LED (NeoPixel) 22
    - 3.9.3. UART TX 23
    - 3.9.4. SPI 23
    - 3.9.5. PWM 25
    - 3.9.6. Using `pioasm` 26
- 4. Using an Integrated Development Environment (IDE) 27
  - 4.1. Using Thonny 27
    - 4.1.1. Blinking the LED from Thonny 28
  - 4.2. Using rshell 29
- Appendix A: App Notes 31
  - Using a SSD1306-based OLED graphics display 31
    - Wiring information 31
    - List of Files 31
    - Bill of Materials 33
  - Using a SH1106-based OLED graphics display 33
    - Wiring information 33
    - List of Files 34
    - Bill of Materials 39
  - Using PIO to drive a set of NeoPixel Ring (WS2812 LEDs) 39
    - Wiring information 39
    - List of Files 40
    - Bill of Materials 41
  - Using UART on the Raspberry Pi Pico 42
    - Wiring information 42
    - List of Files 42
    - Bill of Materials 43
- Appendix B: Documentation Release History 44

Chapter 1. The MicroPython Environment

Python is the fastest way to get started with embedded software on Raspberry Pi Pico. This book is about the official MicroPython port for RP2040-based microcontroller boards.

MicroPython is a Python 3 implementation for microcontrollers and small embedded systems. Because MicroPython is highly efficient, and RP2040 is designed with a disproportionate amount of system memory and processing power for its price, MicroPython is a serious tool for embedded systems development, which does not compromise on approachability.

For exceptionally demanding pieces of software, you can fall back on the SDK (covered in Getting started with Raspberry Pi Pico and Raspberry Pi Pico C/C++ SDK), or an external C module added to your MicroPython firmware, to wring out the very last drop of performance. For every other project, MicroPython handles a lot of heavy lifting for you, and lets you focus on writing the code that adds value to your project. The accelerated floating point libraries in RP2040's on-board ROM storage are used automatically by your Python code, so you should find arithmetic performance quite snappy.

Most on-chip hardware is exposed through the standard `machine` module, so existing MicroPython projects can be ported without too much trouble. The second processor core is exposed through the `_thread` module.

RP2040 has some unique hardware you won't find on other microcontrollers, with the programmable I/O system (PIO) being the prime example of this: a versatile hardware subsystem that lets you create new I/O interfaces and run them at high speed. In the `rp2` module you will find a comprehensive PIO library which lets you write new PIO programs at the MicroPython prompt, and interact with them in real time, to develop interfaces for new or unusual pieces of hardware (or indeed if you just find yourself wanting an extra few serial ports).

MicroPython implements the entire Python 3.4 syntax (including exceptions, `with`, `yield from`, etc., and additionally `async`/`await` keywords from Python 3.5). The following core datatypes are provided: `str` (including basic Unicode support), `bytes`, `bytearray`, `tuple`, `list`, `dict`, `set`, `frozenset`, `array.array`, `collections.namedtuple`, classes and instances. Builtin modules include `sys`, `time`, and `struct`, etc. Note that only a subset of Python 3 functionality is implemented for the data types and modules.

MicroPython can execute scripts in textual source form (`.py` files) or from precompiled bytecode, in both cases either from an on-device filesystem or "frozen" into the MicroPython executable.

1.1. Getting MicroPython for RP2040

> **Pre-built Binary**
>
> A pre-built binary of the latest MicroPython firmware is available from the MicroPython section of the documentation.

The fastest way to get MicroPython is to download the pre-built release binary from the Documentation pages. If you can't or don't want to use the pre-built release — for example, if you want to develop a C module for MicroPython — you can follow the instructions in Section 1.3 to get the source code for MicroPython, which you can use to build your own MicroPython firmware binary.

1.2. Installing MicroPython on Raspberry Pi Pico

Raspberry Pi Pico has a BOOTSEL mode for programming firmware over the USB port. Holding the BOOTSEL button when powering up your board will put it into a special mode where it appears as a USB Mass Storage Device. First make sure your Raspberry Pi Pico is not plugged into *any* source of power: disconnect the micro USB cable if plugged in, and disconnect any other wires that might be providing power to the board, e.g. through the VSYS or VBUS pin. Now hold down the BOOTSEL button, and plug in the micro USB cable (which hopefully has the other end plugged into your computer).

A drive called RPI-RP2 should pop up. Go ahead and drag the MicroPython `firmware.uf2` file onto this drive. This programs the MicroPython firmware onto the flash memory on your Raspberry Pi Pico.

It should take a few seconds to program the UF2 file into the flash. The board will automatically reboot when finished, causing the RPI-RP2 drive to disappear, and boot into MicroPython.

By default, MicroPython doesn't *do* anything when it first boots. It sits and waits for you to type in further instructions. Chapter 2 shows how you can connect with the MicroPython firmware now running on your board. You can read on to see how a custom MicroPython firmware file can be built from the source code.

The Getting started with Raspberry Pi Pico book has detailed instructions on getting your Raspberry Pi Pico into BOOTSEL mode and loading UF2 files, in case you are having trouble. There is also a section going over loading ELF files with the debugger, in case your board doesn't have an easy way of entering BOOTSEL, or you would like to debug a MicroPython C module you are developing.

NOTE

> If you are not following these instructions on a Raspberry Pi Pico, you may not have a BOOTSEL button. If this is the case, you should check if there is some other way of grounding the flash CS pin, such as a jumper, to tell RP2040 to enter the BOOTSEL mode on boot. If there is no such method, you can load code using the Serial Wire Debug interface.

1.3. Building MicroPython From Source

The prebuilt binary which can be downloaded from the MicroPython section of the documentation should serve most use cases, but you can build your own MicroPython firmware from source if you'd like to customise its low-level aspects.

TIP

> If you have already downloaded and installed a prebuilt MicroPython UF2 file, you can skip ahead to Chapter 2 to start using your board.

IMPORTANT

> These instructions for getting and building MicroPython assume you are using Raspberry Pi OS running on a Raspberry Pi 4, or an equivalent Debian-based Linux distribution running on another platform.

It's a good idea to create a pico directory to keep all pico-related checkouts in. These instructions create a pico directory at `/home/pi/pico`.

```
$ cd ~/
$ mkdir pico
$ cd pico
```

Then clone the `micropython` git repository. These instructions will fetch the latest version of the source code.

```
$ git clone https://github.com/micropython/micropython.git --branch master
```

Once the download has finished, the source code for MicroPython should be in a new directory called `micropython`. The MicroPython repository also contains pointers (*submodules*) to specific versions of libraries it needs to run on a particular board, like the SDK in the case of RP2040. We need to fetch these submodules too:

```
$ cd micropython
$ make -C ports/rp2 submodules
```

> **NOTE**
>
> The following instructions assume that you are using a Raspberry Pi Pico. Some details may differ if you are building firmware for a different RP2040-based board. The board vendor should detail any extra steps needed to build firmware for that particular board. The version we're building here is fairly generic, but there might be some differences like putting the default serial port on different pins, or including extra modules to drive that board's hardware.

To build the RP2040 MicroPython port, you'll need to install some extra tools. To build projects you'll need CMake, a cross-platform tool used to build the software, and the GNU Embedded Toolchain for Arm, which turns MicroPython's C source code into a binary program RP2040's processors can understand. `build-essential` is a bundle of tools you need to build code native to your own machine — this is needed for some internal tools in MicroPython and the SDK. You can install all of these via `apt` from the command line. Anything you already have installed will be ignored by `apt`.

```
$ sudo apt update
$ sudo apt install cmake gcc-arm-none-eabi libnewlib-arm-none-eabi build-essential
```

First we need to bootstrap a special tool for MicroPython builds, that ships with the source code:

```
$ make -C mpy-cross
```

We can now build the *port* we need for RP2040, that is, the version of MicroPython that has specific support for our chip.

```
$ cd ports/rp2
$ make
```

If everything went well, there will be a new directory called `build-PICO` (`ports/rp2/build-PICO` relative to the `micropython` directory), which contains the new firmware binaries. The most important ones are:

`firmware.uf2`   A UF2 binary file which can dragged onto the RPI-RP2 drive that pops up once your Raspberry Pi Pico is in BOOTSEL mode. The firmware binary you can download from the documentation page is a UF2 file, because they're the easiest to install.

`firmware.elf` A different type of binary file, which can be loaded by a debugger (such as `gdb` with `openocd`) over RP2040's SWD debug port. This is useful for debugging either a native C module you've added to MicroPython, or the MicroPython core interpreter itself. The actual binary contents is the same as `firmware.uf2`.

You can take a look inside your new `firmware.uf2` using `picotool`, see the Appendix B in the Getting started with Raspberry Pi Pico book for details of how to use `picotool`, e.g.

```
$ picotool info -a build-PICO/firmware.uf2
File build-PICO/firmware.uf2:

Program Information
 name:              MicroPython
 version:           v1.18-412-g965747bd9
 features:          USB REPL
                    thread support
 frozen modules:    _boot, rp2, _boot_fat, ds18x20, onewire, dht, uasyncio,
                    uasyncio/core, uasyncio/event, uasyncio/funcs, uasyncio/lock,
                    uasyncio/stream, neopixel
 binary start:      0x10000000
 binary end:        0x1004ba24
 embedded drive:    0x100a0000-0x10200000 (1408K): MicroPython

Fixed Pin Information
 none

Build Information
 sdk version:       1.3.0
 pico_board:        pico
 boot2_name:        boot2_w25q080
 build date:        May  4 2022
 build attributes:  MinSizeRel
```

Chapter 2. Connecting to the MicroPython REPL

When MicroPython boots for the first time, it will sit and wait for you to connect and tell it what to do. You can load a `.py` file from your computer onto the board, but a more immediate way to interact with it is through what is called the *read-evaluate-print loop*, or REPL (often pronounced similarly to "ripple").

Read MicroPython waits for you to type in some text, followed by the enter key.

Evaluate Whatever you typed is interpreted as Python code, and runs immediately.

Print Any results of the last line you typed are printed out for you to read.

Loop Go back to the start — prompt you for another line of code.

There are two ways to connect to this REPL, so you can communicate with the MicroPython firmware on your board: over USB, and over the UART serial port on Raspberry Pi Pico GPIOs.

2.1. Connecting from a Raspberry Pi over USB

The MicroPython firmware is equipped with a virtual USB serial port which is accessed through the micro USB connector on Raspberry Pi Pico. Your computer should notice this serial port and list it as a character device, most likely `/dev/ttyACM0`.

 TIP

> You can run `ls /dev/tty*` to list your serial ports. There may be quite a few, but MicroPython's USB serial will start with `/dev/ttyACM`. If in doubt, unplug the micro USB connector and see which one disappears. If you don't see anything, you can try rebooting your Raspberry Pi.

You can install `minicom` to access the serial port:

```
$ sudo apt install minicom
```

and then open it as such:

```
$ minicom -o -D /dev/ttyACM0
```

Where the `-D /dev/ttyACM0` is pointing `minicom` at MicroPython's USB serial port, and the `-o` flag essentially means "just do it". There's no need to worry about baud rate, since this is a virtual serial port.

Press the enter key a few times in the terminal where you opened `minicom`. You should see this:

```
>>>
```

This is a *prompt*. MicroPython wants you to type something in, and tell it what to do.

If you press `CTRL-D` on your keyboard whilst the `minicom` terminal is focused, you should see a message similar to this:

```
MPY: soft reboot
MicroPython v1.13-422-g904433073 on 2021-01-19; Raspberry Pi Pico with RP2040
Type "help()" for more information.
>>>
```

This key combination tells MicroPython to reboot. You can do this at any time. When it reboots, MicroPython will print out a message saying exactly what firmware version it is running, and when it was built. Your version number will be different from the one shown here.

2.2. Connecting from a Raspberry Pi using UART

 WARNING

REPL over UART is disabled by default.

The MicroPython port for RP2040 does not expose REPL over a UART port by default. However this default can be changed in the ports/rp2/mpconfigport.h source file. If you want to use the REPL over UART you're going to have to build MicroPython yourself, see Section 1.3 for more details.

Go ahead and download the MicroPython source and in ports/rp2/mpconfigport.h change MICROPY_HW_ENABLE_UART_REPL to 1 to enable it.

```
#define MICROPY_HW_ENABLE_UART_REPL       (1) // useful if there is no USB
```

Then continue to follow the instructions in Section 1.3 to build your own MicroPython UF2 firmware.

This will allow the REPL to be accessed over a UART port, through two GPIO pins. The default settings for UARTs are taken from the C SDK.

Table 1. Default UART settings in MicroPython

Function	Default
UART_BAUDRATE	115,200
UART_BITS	8
UART_STOP	1
UART0_TX	Pin 0
UART0_RX	Pin 1
UART1_TX	Pin 4
UART1_RX	Pin 5

This alternative interface is handy if you have trouble with USB, if you don't have any free USB ports, or if you are using some other RP2040-based board which doesn't have an exposed USB connector.

> **NOTE**
>
> This initially occupies the `UART0` peripheral on RP2040. The `UART1` peripheral is free for you to use in your Python code as a second UART.

The next thing you'll need to do is to enable UART serial on the Raspberry Pi. To do so, run `raspi-config`,

```
$ sudo raspi-config
```

and go to `Interfacing Options` → `Serial` and select "No" when asked "Would you like a login shell to be accessible over serial?" and "Yes" when asked "Would you like the serial port hardware to be enabled?". You should see something like Figure 1.

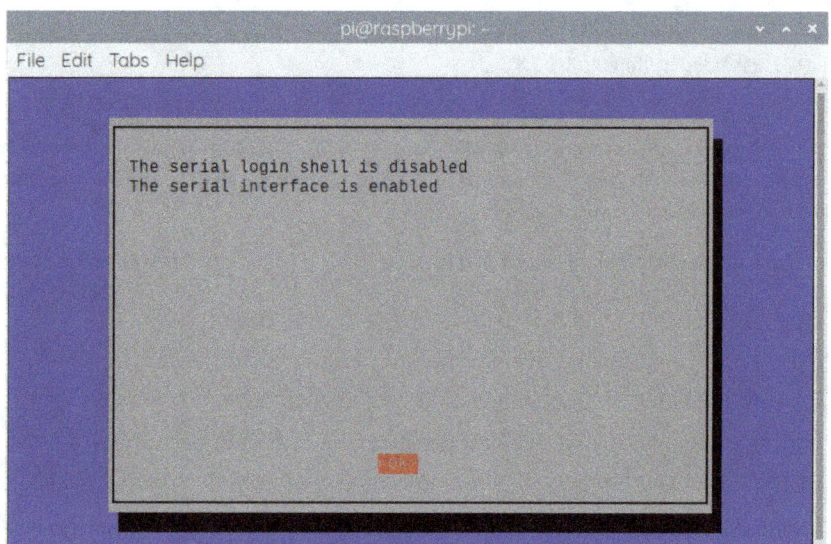

Figure 1. Enabling a serial UART using `raspi-config` on the Raspberry Pi.

Leaving `raspi-config` you should choose "Yes" and reboot your Raspberry Pi to enable the serial port.

You should then wire the Raspberry Pi and the Raspberry Pi Pico together with the following mapping:

Raspberry Pi	Raspberry Pi Pico
GND	GND
GPIO15 (UART_RX0)	GPIO0 (UART0_TX)
GPIO14 (UART_TX0)	GPOI1 (UART0_RX)

> **IMPORTANT**
>
> RX matches to TX, and TX matches to RX. You mustn't connect the two opposite TX pins together, or the two RX pins. This is because MicroPython needs to listen on the channel that the Raspberry Pi transmits on, and vice versa.

See Figure 2.

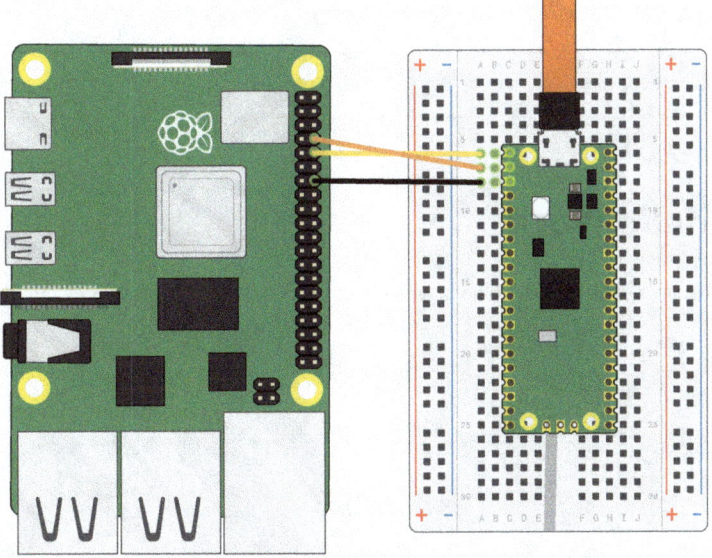

Figure 2. A Raspberry Pi 4 and the Raspberry Pi Pico with UART0 connected together.

then connect to the board using `minicom` connected to `/dev/serial0`,

```
$ minicom -b 115200 -o -D /dev/serial0
```

If you press the enter key, MicroPython should respond by prompting you for more input:

```
>>>
```

2.3. Connecting from a Mac

So long as you're using a recent version of macOS like Catalina, drivers should already be loaded. Otherwise see the manufacturers' website for FTDI Chip Drivers. Then you should use a Terminal program to connect to Serial-over-USB (USB CDC). The serial port will show up as `/dev/tty.usbmodem0000000000001`

If you don't already have a Terminal program installed you can install `minicom` using Homebrew,

```
$ brew install minicom
```

and connect to the board as below.

```
$ minicom -b 115200 -o -D /dev/tty.usbmodem0000000000001
```

> **ℹ NOTE**
>
> Other Terminal applications like CoolTerm or Serial can also be used.

2.4. Say "Hello World"

Once connected you can check that everything is working by typing a Python "Hello World" into the REPL,

```
>>> print("Hello, Pico!")
Hello, Pico!
>>>
```

2.5. Blink an LED

The on-board LED on Raspberry Pi Pico is connected to GPIO pin 25. You can blink this on and off from the REPL. When you see the REPL prompt enter the following,

```
>>> from machine import Pin
>>> led = Pin(25, Pin.OUT)
```

The `machine` module is used to control on-chip hardware. This is standard on all MicroPython ports, and you can read more about it in the MicroPython documentation. Here we are using it to take control of a GPIO, so we can drive it high and low. If you type this in,

```
>>> led.value(1)
```

The LED should turn on. You can turn it off again with

```
>>> led.value(0)
```

2.6. What next?

At this point you should have MicroPython installed on your board, and have tested your setup by typing short programs into the prompt to print some text back to you, and blink an LED.

You can read on to the next chapter, which goes into the specifics of MicroPython on RP2040, and where it differs from other platforms. Chapter 3 also has some short examples of the different APIs offered to interact with the hardware.

You can learn how to set up an *integrated development environment* (IDE) in Chapter 4, so you don't have to type programs in line by line.

You can dive straight into Appendix A if you are eager to start connecting wires to a breadboard.

Chapter 3. The RP2040 Port

Currently supported features include:

- REPL over USB and UART (on GP0/GP1).
- 1600kB filesystem using `littlefs2` on the on-board flash. (Default size for Raspberry Pi Pico)
- `utime` module with sleep and ticks functions.
- `ubinascii` module.
- `machine` module with some basic functions.
    - `machine.Pin` class.
    - `machine.Timer` class.
    - `machine.ADC` class.
    - `machine.I2C` and `machine.SoftI2C` classes.
    - `machine.SPI` and `machine.SoftSPI` classes.
    - `machine.WDT` class.
    - `machine.PWM` class.
    - `machine.UART` class.
- `rp2` platform-specific module.
    - PIO hardware access library
    - PIO program assembler
    - Raw flash read/write access
- Multicore support exposed via the standard `_thread` module
- Accelerated floating point arithmetic using the RP2040 ROM library and hardware divider (used automatically)

Documentation around MicroPython is available from https://docs.micropython.org. For example the `machine` module, which can be used to access a lot of RP2040's on-chip hardware, is standard, and you will find a lot of the information you need in the online documentation for that module.

This chapter will give a very brief tour of some of the hardware APIs, with code examples you can either type into the REPL (Chapter 2) or load onto the board using a development environment installed on your computer (Chapter 4).

3.1. Blinking an LED Forever (Timer)

In Chapter 2 we saw how the `machine.Pin` class could be used to turn an LED on and off, by driving a GPIO high and low.

```
>>> from machine import Pin
>>> led = Pin(25, Pin.OUT)
>>> led.value(1)
>>> led.value(0)
```

This is, to put it mildy, quite a convoluted way of turning a light on and off. A light switch would work better. The `machine.Timer` class, which uses RP2040's hardware timer to trigger callbacks at regular intervals, saves a lot of typing if we want the light to turn itself on and off repeatedly, thus bringing our level of automation from "mechanical switch" to "555 timer".

Pico MicroPython Examples: https://github.com/raspberrypi/pico-micropython-examples/blob/master/blink/blink.py

```python
from machine import Pin, Timer

led = Pin("LED", Pin.OUT)
tim = Timer()
def tick(timer):
    global led
    led.toggle()

tim.init(freq=2.5, mode=Timer.PERIODIC, callback=tick)
```

Typing this program into the REPL will cause the LED to start blinking, but the prompt will appear again:

```
>>>
```

The `Timer` we created will run in the background, at the interval we specified, blinking the LED. The MicroPython prompt is still running in the foreground, and we can enter more code, or start more timers.

3.2. UART

NOTE

REPL over UART is disabled by default. See Section 2.2 for details of how to enable REPL over UART.

Example usage looping UART0 to UART1.

Pico MicroPython Examples: https://github.com/raspberrypi/pico-micropython-examples/blob/master/uart/loopback/uart.py

```python
from machine import UART, Pin
import time

uart1 = UART(1, baudrate=9600, tx=Pin(8), rx=Pin(9))

uart0 = UART(0, baudrate=9600, tx=Pin(0), rx=Pin(1))

txData = b'hello world\n\r'
uart1.write(txData)
time.sleep(0.1)
rxData = bytes()
while uart0.any() > 0:
    rxData += uart0.read(1)

print(rxData.decode('utf-8'))
```

For more detail, including a wiring diagram, see Appendix A.

3.3. ADC

An analogue-to-digital converter (ADC) measures some analogue signal and encodes it as a digital number. The ADC on RP2040 measures voltages.

An ADC has two key features: its resolution, measured in digital bits, and its channels, or how many analogue signals it can accept and convert at once. The ADC on RP2040 has a resolution of 12-bits, meaning that it can transform an analogue signal into a digital signal as a number ranging from 0 to 4095 – though this is handled in MicroPython transformed to a 16-bit number ranging from 0 to 65,535, so that it behaves the same as the ADC on other MicroPython microcontrollers.

RP2040 has five ADC channels total, four of which are brought out to chip GPIOs: GP26, GP27, GP28 and GP29. On Raspberry Pi Pico, the first three of these are brought out to GPIO pins, and the fourth can be used to measure the VSYS voltage on the board.

The ADC's fifth input channel is connected to a temperature sensor built into RP2040.

You can specify which ADC channel you're using by pin number, e.g.

```python
adc = machine.ADC(26) # Connect to GP26, which is channel 0
```

or by channel,

```python
adc = machine.ADC(4) # Connect to the internal temperature sensor
adc = machine.ADC(0) # Connect to channel 0 (GP26)
```

An example reading the fourth analogue-to-digital (ADC) converter channel, connected to the internal temperature sensor:

Pico MicroPython Examples: https://github.com/raspberrypi/pico-micropython-examples/blob/master/adc/temperature.py

```python
import machine
import utime

sensor_temp = machine.ADC(4)
conversion_factor = 3.3 / (65535)

while True:
    reading = sensor_temp.read_u16() * conversion_factor

    # The temperature sensor measures the Vbe voltage of a biased bipolar diode, connected to the fifth ADC channel
    # Typically, Vbe = 0.706V at 27 degrees C, with a slope of -1.721mV (0.001721) per degree.

    temperature = 27 - (reading - 0.706)/0.001721
    print(temperature)
    utime.sleep(2)
```

3.4. Interrupts

You can set an IRQ like this:

Pico MicroPython Examples: https://github.com/raspberrypi/pico-micropython-examples/blob/master/irq/irq.py

```python
from machine import Pin

p2 = Pin(2, Pin.IN, Pin.PULL_UP)
p2.irq(lambda pin: print("IRQ with flags:", pin.irq().flags()), Pin.IRQ_FALLING)
```

It should print out something when GP2 has a falling edge.

3.5. Multicore Support

Example usage:

Pico MicroPython Examples: https://github.com/raspberrypi/pico-micropython-examples/blob/master/multicore/multicore.py

```python
import time, _thread, machine

def task(n, delay):
    led = machine.Pin(25, machine.Pin.OUT)
    for i in range(n):
        led.high()
        time.sleep(delay)
        led.low()
        time.sleep(delay)
    print('done')

_thread.start_new_thread(task, (10, 0.5))
```

Only one thread can be started/running at any one time, because there is no `RTOS` just a second core. The `GIL` is not enabled so both `core0` and `core1` can run Python code concurrently, with care to use locks for shared data.

3.6. I2C

Example usage:

Pico MicroPython Examples: https://github.com/raspberrypi/pico-micropython-examples/blob/master/i2c/i2c.py

```python
from machine import Pin, I2C

i2c = I2C(0, scl=Pin(9), sda=Pin(8), freq=100000)
i2c.scan()
i2c.writeto(76, b'123')
i2c.readfrom(76, 4)


i2c = I2C(1, scl=Pin(7), sda=Pin(6), freq=100000)
i2c.scan()
i2c.writeto_mem(76, 6, b'456')
i2c.readfrom_mem(76, 6, 4)
```

I2C can be constructed without specifying the frequency, if you just want all the defaults.

Pico MicroPython Examples: https://github.com/raspberrypi/pico-micropython-examples/blob/master/i2c/i2c_without_freq.py

```python
from machine import I2C

i2c = I2C(0)   # defaults to SCL=Pin(9), SDA=Pin(8), freq=400000
```

> ⛔ **WARNING**
>
> There may be some bugs reading/writing to device addresses that do not respond, the hardware seems to lock up in some cases.

Table 2. Default I2C pins

Function	Default
I2C Frequency	400,000
I2C0 SCL	Pin 9
I2C0 SDA	Pin 8
I2C1 SCL	Pin 7
I2C1 SDA	Pin 6

3.7. SPI

Example usage:

Pico MicroPython Examples: https://github.com/raspberrypi/pico-micropython-examples/blob/master/spi/spi.py

```python
from machine import SPI

spi = SPI(0)
spi = SPI(0, 100_000)
spi = SPI(0, 100_000, polarity=1, phase=1)

spi.write('test')
spi.read(5)

buf = bytearray(3)
spi.write_readinto('out', buf)
```

> ℹ️ **NOTE**
>
> The chip select must be managed separately using a `machine.Pin`.

Table 3. Default SPI pins

Function	Default
SPI_BAUDRATE	1,000,000
SPI_POLARITY	0
SPI_PHASE	0
SPI_BITS	8
SPI_FIRSTBIT	MSB
SPI0_SCK	Pin 6
SPI0_MOSI	Pin 7
SPI0_MISO	Pin 4
SPI1_SCK	Pin 10
SPI1_MOSI	Pin 11

SPI1_MISO	Pin 8

3.8. PWM

Example of using PWM to fade an LED:

Pico MicroPython Examples: https://github.com/raspberrypi/pico-micropython-examples/blob/master/pwm/pwm_fade.py

```python
# Example using PWM to fade an LED.

import time
from machine import Pin, PWM


# Construct PWM object, with LED on Pin(25).
pwm = PWM(Pin(25))

# Set the PWM frequency.
pwm.freq(1000)

# Fade the LED in and out a few times.
duty = 0
direction = 1
for _ in range(8 * 256):
    duty += direction
    if duty > 255:
        duty = 255
        direction = -1
    elif duty < 0:
        duty = 0
        direction = 1
    pwm.duty_u16(duty * duty)
    time.sleep(0.001)
```

3.9. PIO Support

Current support allows you to define Programmable IO (PIO) Assembler blocks and using them in the PIO peripheral, more documentation around PIO can be found in Chapter 3 of the RP2040 Datasheet and Chapter 4 of the Raspberry Pi Pico C/C++ SDK book.

The Raspberry Pi Pico MicroPython introduces a new `@rp2.asm_pio` decorator, along with a `rp2.PIO` class. The definition of a PIO program, and the configuration of the state machine, into 2 logical parts:

- The program definition, including how many pins are used and if they are in/out pins. This goes in the `@rp2.asm_pio` definition. This is close to what the `pioasm` tool from the SDK would generate from a `.pio` file (but here it's all defined in Python).

- The program instantiation, which sets the frequency of the state machine and which pins to bind to. These get set when setting a SM to run a particular program.

The aim was to allow a program to be defined once and then easily instantiated multiple times (if needed) with different GPIO. Another aim was to make it easy to do basic things without getting weighed down in too much PIO/SM configuration.

Example usage, to blink the on-board LED connected to GPIO 25,

Pico MicroPython Examples: *https://github.com/raspberrypi/pico-micropython-examples/blob/master/pio/pio_blink.py*

```
 1  import time
 2  import rp2
 3  from machine import Pin
 4
 5  # Define the blink program.  It has one GPIO to bind to on the set instruction, which is an
    output pin.
 6  # Use lots of delays to make the blinking visible by eye.
 7  @rp2.asm_pio(set_init=rp2.PIO_OUT_LOW)
 8  def blink():
 9      wrap_target()
10      set(pins, 1)   [31]
11      nop()          [31]
12      nop()          [31]
13      nop()          [31]
14      nop()          [31]
15      set(pins, 0)   [31]
16      nop()          [31]
17      nop()          [31]
18      nop()          [31]
19      nop()          [31]
20      wrap()
21
22  # Instantiate a state machine with the blink program, at 2000Hz, with set bound to Pin(25)
    (LED on the rp2 board)
23  sm = rp2.StateMachine(0, blink, freq=2000, set_base=Pin(25))
24
25  # Run the state machine for 3 seconds.  The LED should blink.
26  sm.active(1)
27  time.sleep(3)
28  sm.active(0)
```

or via explicit exec.

Pico MicroPython Examples: *https://github.com/raspberrypi/pico-micropython-examples/blob/master/pio/pio_exec.py*

```
 1  # Example using PIO to turn on an LED via an explicit exec.
 2  #
 3  # Demonstrates:
 4  #   - using set_init and set_base
 5  #   - using StateMachine.exec
 6
 7  import time
 8  from machine import Pin
 9  import rp2
10
11  # Define an empty program that uses a single set pin.
12  @rp2.asm_pio(set_init=rp2.PIO_OUT_LOW)
13  def prog():
14      pass
15
16
17  # Construct the StateMachine, binding Pin(25) to the set pin.
18  sm = rp2.StateMachine(0, prog, set_base=Pin(25))
19
20  # Turn on the set pin via an exec instruction.
21  sm.exec("set(pins, 1)")
22
23  # Sleep for 500ms.
24  time.sleep(0.5)
```

```
25
26 # Turn off the set pin via an exec instruction.
27 sm.exec("set(pins, 0)")
```

Some points to note,

- All program configuration (eg autopull) is done in the `@asm_pio` decorator. Only the frequency and base pins are set in the StateMachine constructor.
- `[n]` is used for delay, `.set(n)` used for sideset
- The assembler will automatically detect if sideset is used everywhere or only on a few instructions, and set the `SIDE_EN` bit automatically

The idea is that for the 4 sets of pins (`in`, `out`, `set`, `sideset`, excluding `jmp`) that can be connected to a state machine, there's the following that need configuring for each set:

1. base GPIO
2. number of consecutive GPIO
3. initial GPIO direction (in or out pin)
4. initial GPIO value (high or low)

In the design of the Python API for PIO these 4 items are split into "declaration" (items 2-4) and "instantiation" (item 1). In other words, a program is written with items 2-4 fixed for that program (eg a WS2812 driver would have 1 output pin) and item 1 is free to change without changing the program (eg which pin the WS2812 is connected to).

So in the `@asm_pio` decorator you declare items 2-4, and in the `StateMachine` constructor you say which base pin to use (item 1). That makes it easy to define a single program and instantiate it multiple times on different pins (you can't really change items 2-4 for a different instantiation of the same program, it doesn't really make sense to do that).

And the same keyword arg (in the case about it's `sideset_pins`) is used for both the declaration and instantiation, to show that they are linked.

To declare multiple pins in the decorator (the count, ie item 2 above), you use a tuple/list of values. And each item in the tuple/list specified items 3 and 4. For example:

```
1 @asm_pio(set_pins=(PIO.OUT_LOW, PIO.OUT_HIGH, PIO.IN_LOW), sideset_pins=PIO.OUT_LOW)
2 def foo():
3     ....
4
5 sm = StateMachine(0, foo, freq=10000, set_pins=Pin(15), sideset_pins=Pin(22))
```

In this example:

- there are 3 set pins connected to the SM, and their initial state (set when the StateMachine is created) is: output low, output high, input low (used for open-drain)
- there is 1 sideset pin, initial state is output low
- the 3 set pins start at Pin(15)
- the 1 sideset pin starts at Pin(22)

The reason to have the constants `OUT_LOW`, `OUT_HIGH`, `IN_LOW` and `IN_HIGH` is so that the pin value and dir are automatically set before the start of the PIO program (instead of wasting instruction words to do `set(pindirs, 1)` etc at the start).

3.9.1. IRQ

There is support for PIO IRQs, e.g.

Pico MicroPython Examples: https://github.com/raspberrypi/pico-micropython-examples/blob/master/pio/pio_irq.py

```python
import time
import rp2

@rp2.asm_pio()
def irq_test():
    wrap_target()
    nop()           [31]
    nop()           [31]
    nop()           [31]
    nop()           [31]
    irq(0)
    nop()           [31]
    nop()           [31]
    nop()           [31]
    nop()           [31]
    irq(1)
    wrap()


rp2.PIO(0).irq(lambda pio: print(pio.irq().flags()))

sm = rp2.StateMachine(0, irq_test, freq=2000)
sm.active(1)
time.sleep(1)
sm.active(0)
```

An example program that blinks at 1Hz and raises an IRQ at 1Hz to print the current millisecond timestamp,

Pico MicroPython Examples: https://github.com/raspberrypi/pico-micropython-examples/blob/master/pio/pio_1hz.py

```python
# Example using PIO to blink an LED and raise an IRQ at 1Hz.

import time
from machine import Pin
import rp2


@rp2.asm_pio(set_init=rp2.PIO.OUT_LOW)
def blink_1hz():
    # Cycles: 1 + 1 + 6 + 32 * (30 + 1) = 1000
    irq(rel(0))
    set(pins, 1)
    set(x, 31)                  [5]
    label("delay_high")
    nop()                       [29]
    jmp(x_dec, "delay_high")

    # Cycles: 1 + 7 + 32 * (30 + 1) = 1000
    set(pins, 0)
    set(x, 31)                  [6]
    label("delay_low")
    nop()                       [29]
    jmp(x_dec, "delay_low")


# Create the StateMachine with the blink_1hz program, outputting on Pin(25).
sm = rp2.StateMachine(0, blink_1hz, freq=2000, set_base=Pin(25))

# Set the IRQ handler to print the millisecond timestamp.
```

```
30 sm.irq(lambda p: print(time.ticks_ms()))
31
32 # Start the StateMachine.
33 sm.active(1)
```

or to wait for a pin change and raise an IRQ.

Pico MicroPython Examples: *https://github.com/raspberrypi/pico-micropython-examples/blob/master/pio/pio_pinchange.py*

```
1  # Example using PIO to wait for a pin change and raise an IRQ.
2  #
3  # Demonstrates:
4  #   - PIO wrapping
5  #   - PIO wait instruction, waiting on an input pin
6  #   - PIO irq instruction, in blocking mode with relative IRQ number
7  #   - setting the in_base pin for a StateMachine
8  #   - setting an irq handler for a StateMachine
9  #   - instantiating 2x StateMachine's with the same program and different pins
10
11 import time
12 from machine import Pin
13 import rp2
14
15
16 @rp2.asm_pio()
17 def wait_pin_low():
18     wrap_target()
19
20     wait(0, pin, 0)
21     irq(block, rel(0))
22     wait(1, pin, 0)
23
24     wrap()
25
26
27 def handler(sm):
28     # Print a (wrapping) timestamp, and the state machine object.
29     print(time.ticks_ms(), sm)
30
31
32 # Instantiate StateMachine(0) with wait_pin_low program on Pin(16).
33 pin16 = Pin(16, Pin.IN, Pin.PULL_UP)
34 sm0 = rp2.StateMachine(0, wait_pin_low, in_base=pin16)
35 sm0.irq(handler)
36
37 # Instantiate StateMachine(1) with wait_pin_low program on Pin(17).
38 pin17 = Pin(17, Pin.IN, Pin.PULL_UP)
39 sm1 = rp2.StateMachine(1, wait_pin_low, in_base=pin17)
40 sm1.irq(handler)
41
42 # Start the StateMachine's running.
43 sm0.active(1)
44 sm1.active(1)
45
46 # Now, when Pin(16) or Pin(17) is pulled low a message will be printed to the REPL.
```

3.9.2. WS2812 LED (NeoPixel)

While a WS2812 LED (NeoPixel) can be driven via the following program,

Pico MicroPython Examples: https://github.com/raspberrypi/pico-micropython-examples/blob/master/pio/pio_ws2812.py

```python
# Example using PIO to drive a set of WS2812 LEDs.

import array, time
from machine import Pin
import rp2

# Configure the number of WS2812 LEDs.
NUM_LEDS = 8


@rp2.asm_pio(sideset_init=rp2.PIO.OUT_LOW, out_shiftdir=rp2.PIO.SHIFT_LEFT, autopull=True, pull_thresh=24)
def ws2812():
    T1 = 2
    T2 = 5
    T3 = 3
    wrap_target()
    label("bitloop")
    out(x, 1)               .side(0)    [T3 - 1]
    jmp(not_x, "do_zero")   .side(1)    [T1 - 1]
    jmp("bitloop")          .side(1)    [T2 - 1]
    label("do_zero")
    nop()                   .side(0)    [T2 - 1]
    wrap()


# Create the StateMachine with the ws2812 program, outputting on Pin(22).
sm = rp2.StateMachine(0, ws2812, freq=8_000_000, sideset_base=Pin(22))

# Start the StateMachine, it will wait for data on its FIFO.
sm.active(1)

# Display a pattern on the LEDs via an array of LED RGB values.
ar = array.array("I", [0 for _ in range(NUM_LEDS)])

# Cycle colours.
for i in range(4 * NUM_LEDS):
    for j in range(NUM_LEDS):
        r = j * 100 // (NUM_LEDS - 1)
        b = 100 - j * 100 // (NUM_LEDS - 1)
        if j != i % NUM_LEDS:
            r >>= 3
            b >>= 3
        ar[j] = r << 16 | b
    sm.put(ar, 8)
    time.sleep_ms(50)

# Fade out.
for i in range(24):
    for j in range(NUM_LEDS):
        ar[j] >>= 1
    sm.put(ar, 8)
    time.sleep_ms(50)
```

3.9.3. UART TX

A UART TX example,

Pico MicroPython Examples: https://github.com/raspberrypi/pico-micropython-examples/blob/master/pio/pio_uart_tx.py

```python
# Example using PIO to create a UART TX interface

from machine import Pin
from rp2 import PIO, StateMachine, asm_pio

UART_BAUD = 115200
PIN_BASE = 10
NUM_UARTS = 8


@asm_pio(sideset_init=PIO.OUT_HIGH, out_init=PIO.OUT_HIGH, out_shiftdir=PIO.SHIFT_RIGHT)
def uart_tx():
    # Block with TX deasserted until data available
    pull()
    # Initialise bit counter, assert start bit for 8 cycles
    set(x, 7)   .side(0)       [7]
    # Shift out 8 data bits, 8 execution cycles per bit
    label("bitloop")
    out(pins, 1)               [6]
    jmp(x_dec, "bitloop")
    # Assert stop bit for 8 cycles total (incl 1 for pull())
    nop()       .side(1)       [6]



# Now we add 8 UART TXs, on pins 10 to 17. Use the same baud rate for all of them.
uarts = []
for i in range(NUM_UARTS):
    sm = StateMachine(
        i, uart_tx, freq=8 * UART_BAUD, sideset_base=Pin(PIN_BASE + i), out_base=Pin(PIN_BASE + i)
    )
    sm.active(1)
    uarts.append(sm)

# We can print characters from each UART by pushing them to the TX FIFO
def pio_uart_print(sm, s):
    for c in s:
        sm.put(ord(c))



# Print a different message from each UART
for i, u in enumerate(uarts):
    pio_uart_print(u, "Hello from UART {}!\n".format(i))
```

> **NOTE**
>
> You need to specify an initial OUT pin state in your program in order to be able to pass OUT mapping to your SM instantiation, even though in this program it is redundant because the mappings overlap.

3.9.4. SPI

An SPI example.

Pico MicroPython Examples: https://github.com/raspberrypi/pico-micropython-examples/blob/master/pio/pio_spi.py

```python
import rp2
from machine import Pin

@rp2.asm_pio(out_shiftdir=0, autopull=True, pull_thresh=8, autopush=True, push_thresh=8,
    sideset_init=(rp2.PIO.OUT_LOW, rp2.PIO.OUT_HIGH), out_init=rp2.PIO.OUT_LOW)
def spi_cpha0():
    # Note X must be preinitialised by setup code before first byte, we reload after sending each byte
    # Would normally do this via exec() but in this case it's in the instruction memory and is only run once
    set(x, 6)
    # Actual program body follows
    wrap_target()
    pull(ifempty)           .side(0x2)  [1]
    label("bitloop")
    out(pins, 1)            .side(0x0)  [1]
    in_(pins, 1)            .side(0x1)
    jmp(x_dec, "bitloop")   .side(0x1)

    out(pins, 1)            .side(0x0)
    set(x, 6)               .side(0x0) # Note this could be replaced with mov x, y for programmable frame size
    in_(pins, 1)            .side(0x1)
    jmp(not_osre, "bitloop") .side(0x1) # Fallthru if TXF empties

    nop()                   .side(0x0)  [1] # CSn back porch
    wrap()


class PIOSPI:

    def __init__(self, sm_id, pin_mosi, pin_miso, pin_sck, cpha=False, cpol=False, freq=1000000):
        assert(not(cpol or cpha))
        self._sm = rp2.StateMachine(sm_id, spi_cpha0, freq=4*freq, sideset_base=Pin(pin_sck), out_base=Pin(pin_mosi), in_base=Pin(pin_sck))
        self._sm.active(1)

    # Note this code will die spectacularly cause we're not draining the RX FIFO
    def write_blocking(wdata):
        for b in wdata:
            self._sm.put(b << 24)

    def read_blocking(n):
        data = []
        for i in range(n):
            data.append(self._sm.get() & 0xff)
        return data

    def write_read_blocking(wdata):
        rdata = []
        for b in wdata:
            self._sm.put(b << 24)
            rdata.append(self._sm.get() & 0xff)
        return rdata
```

> **NOTE**
>
> This SPI program supports programmable frame sizes (by holding the reload value for X counter in the Y register) but currently this can't be used, because the autopull threshold is associated with the program, instead of the SM instantiation.

3.9.5. PWM

A PWM example,

Pico MicroPython Examples: https://github.com/raspberrypi/pico-micropython-examples/blob/master/pio/pio_pwm.py

```python
# Example of using PIO for PWM, and fading the brightness of an LED

from machine import Pin
from rp2 import PIO, StateMachine, asm_pio
from time import sleep


@asm_pio(sideset_init=PIO.OUT_LOW)
def pwm_prog():
    pull(noblock) .side(0)
    mov(x, osr) # Keep most recent pull data stashed in X, for recycling by noblock
    mov(y, isr) # ISR must be preloaded with PWM count max
    label("pwmloop")
    jmp(x_not_y, "skip")
    nop()         .side(1)
    label("skip")
    jmp(y_dec, "pwmloop")


class PIOPWM:
    def __init__(self, sm_id, pin, max_count, count_freq):
        self._sm = StateMachine(sm_id, pwm_prog, freq=2 * count_freq, sideset_base=Pin(pin))
        # Use exec() to load max count into ISR
        self._sm.put(max_count)
        self._sm.exec("pull()")
        self._sm.exec("mov(isr, osr)")
        self._sm.active(1)
        self._max_count = max_count

    def set(self, value):
        # Minimum value is -1 (completely turn off), 0 actually still produces narrow pulse
        value = max(value, -1)
        value = min(value, self._max_count)
        self._sm.put(value)


# Pin 25 is LED on Pico boards
pwm = PIOPWM(0, 25, max_count=(1 << 16) - 1, count_freq=10_000_000)

while True:
    for i in range(256):
        pwm.set(i ** 2)
        sleep(0.01)
```

3.9.6. Using `pioasm`

As well as writing PIO code inline in your MicroPython script you can use the `pioasm` tool from the C/C++ SDK to generate a Python file.

```
$ pioasm -o python input (output)
```

For more information on `pioasm` see the **Raspberry Pi Pico C/C++ SDK** book which talks about the C/C++ SDK.

Chapter 4. Using an Integrated Development Environment (IDE)

The MicroPython port to Raspberry Pi Pico and other RP2040-based boards works with commonly used development environments.

4.1. Using Thonny

Thonny packages are available for Linux, MS Windows, and macOS. After installation, using the Thonny development environment is the same across all three platforms. The latest release of Thonny can be downloaded from thonny.org

Alternatively if you are working on a Raspberry Pi you should install Thonny using apt from the command line,

```
$ sudo apt install thonny
```

this will add a Thonny icon to the Raspberry Pi desktop menu. Go ahead and select Raspberry Pi → Programming → Thonny Python IDE to open the development environment.

When opening Thonny for the first time select "Standard Mode." For some versions this choice will be made via a popup when you first open Thonny. However for the Raspberry Pi release you should click on the text in the top right of the window to switch to "Regular Mode."

Make sure your Raspberry Pi Pico is plugged into your computer and, click on the word 'Python' followed by a version number at the bottom-right of the Thonny window — this is the Python interpreter that Thonny is currently using. Normally the interpreter is the copy of Python running on Raspberry Pi, but it needs to be changed in order to run your programs in MicroPython on your Pico, clicking the current interpreter will open a drop down.

Select "MicroPython (Raspberry Pi Pico)" from the list, see Figure 3.

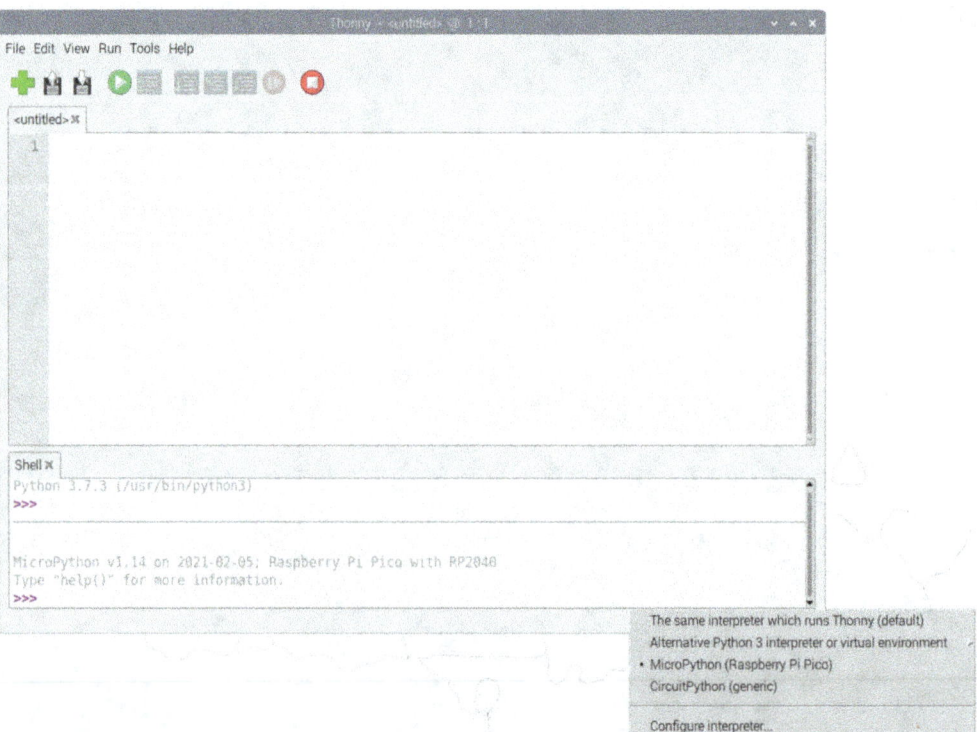

Figure 3. Switching to MicroPython

> **NOTE**
>
> The Raspberry Pi Pico interpreter is only available in the latest version of Thonny. If you're running an older version and can't update it, look for 'MicroPython (generic)' instead. If your version of Thonny is older still and has no interpreter option at the bottom-right of the window and you can't update it, restart Thonny, click the "Run" menu, and click 'Select interpreter.' Click the drop-down arrow next to 'The same interpreter that runs Thonny (default)', click on 'MicroPython (generic)' in the list, then click on the drop-down arrow next to 'Port' and click on 'Board in FS mode' in that list before clicking "OK" to confirm your changes.

You can now access the REPL from the Shell panel,

```
>>> print('Hello Pico!')
Hello Pico!
>>>
```

see Figure 4.

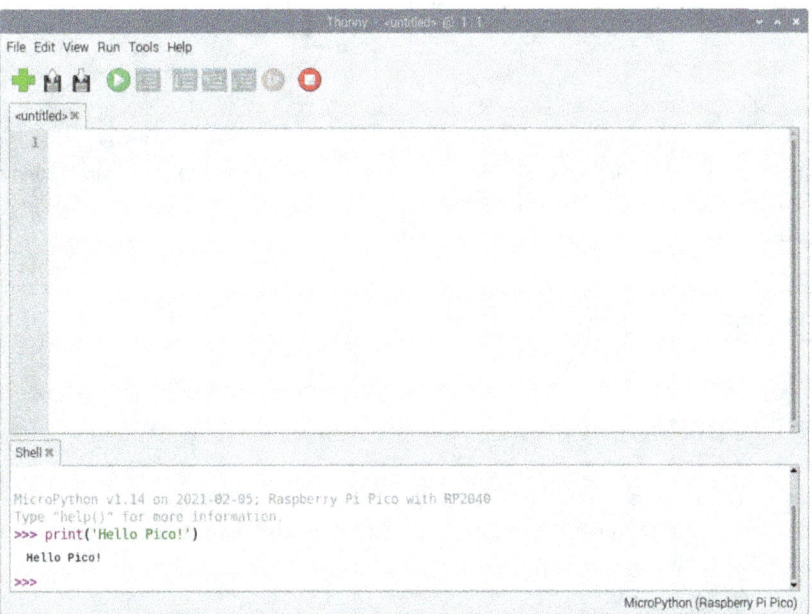

Figure 4. Saying "Hello Pico!" from the MicroPython REPL inside the Thonny environment.

4.1.1. Blinking the LED from Thonny

You can use a timer to blink the on-board LED.

Pico MicroPython Examples: https://github.com/raspberrypi/pico-micropython-examples/blob/master/blink/blink.py

```python
1  from machine import Pin, Timer
2
3  led = Pin("LED", Pin.OUT)
4  tim = Timer()
5  def tick(timer):
6      global led
7      led.toggle()
8
9  tim.init(freq=2.5, mode=Timer.PERIODIC, callback=tick)
```

Enter the code in the main panel, then click on the green run button. Thonny will present you with a popup, click on

"MicroPython device" and enter "test.py" to save the code to the Raspberry Pi Pico, see Figure 5.

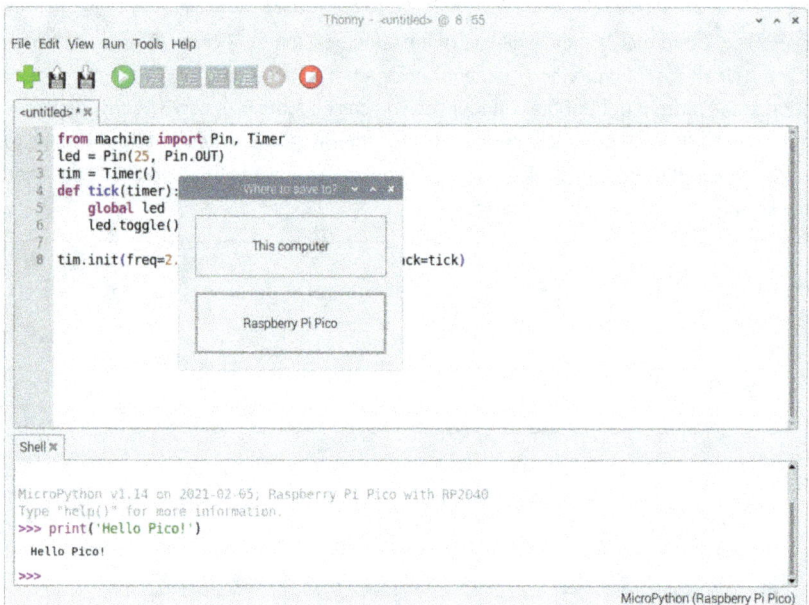

Figure 5. Saving code to the Raspberry Pi Pico inside the Thonny environment.

> **NOTE**
>
> If you "save a file to the device" and give it the special name `main.py`, then MicroPython starts running that script as soon as power is supplied to Raspberry Pi Pico in the future.

The program should uploaded to the Raspberry Pi Pico using the REPL, and automatically start running. You should see the on-board LED start blinking, connected to GPIO pin 25

4.2. Using rshell

The Remote Shell for MicroPython (`rshell`) is a simple shell which runs on the host and uses MicroPython's REPL to send python code to the Raspberry Pi Pico in order to get filesystem information, and to copy files to and from MicroPython's own filesystem.

You can install rshell using,

```
$ sudo apt install python3-pip
$ sudo pip3 install rshell
```

You can then connect to Raspberry Pi Pico using,

```
$ rshell --buffer-size=512 -p /dev/ttyACM0
Connecting to /dev/ttyACM0 (buffer-size 512)...
Trying to connect to REPL  connected
Testing if sys.stdin.buffer exists ... N
Retrieving root directories ...
Setting time ... Aug 21, 2020 15:35:18
Evaluating board_name ... pyboard
Retrieving time epoch ... Jan 01, 2000
Welcome to rshell. Use Control-D (or the exit command) to exit rshell.
/home/pi>
```

Full documentation of `rshell` can be found on the project's GitHub repository.

Appendix A: App Notes

Using a SSD1306-based OLED graphics display

Display an image and text on I2C driven SSD1306-based OLED graphics display.

Wiring information

See Figure 6 for wiring instructions.

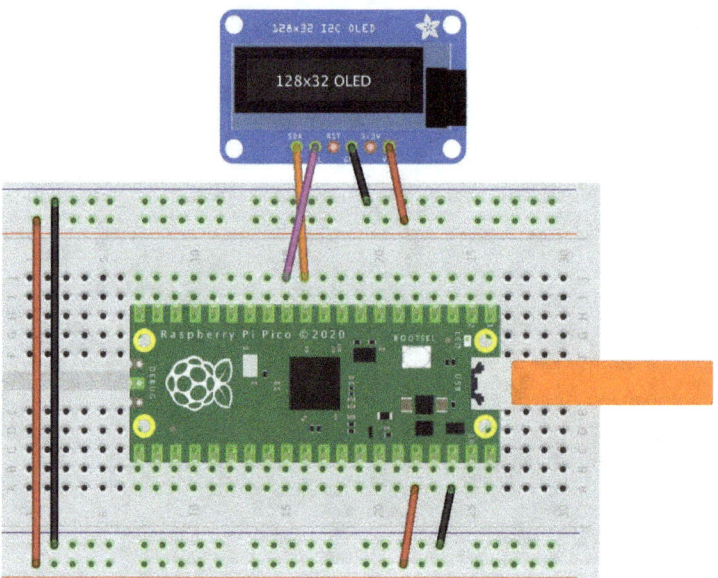

Figure 6. Wiring the OLED to Pico using I2C

List of Files

A list of files with descriptions of their function;

i2c_1306oled_using_defaults.py

  The example code.

Pico MicroPython Examples: https://github.com/raspberrypi/pico-micropython-examples/blob/master/i2c/1306oled/i2c_1306oled_using_defaults.py

```python
1  # Display Image & text on I2C driven ssd1306 OLED display
2  from machine import Pin, I2C
3  from ssd1306 import SSD1306_I2C
4  import framebuf
5
6  WIDTH = 128                                              # oled display width
7  HEIGHT = 32                                              # oled display height
8
9  i2c = I2C(0)                                             # Init I2C using I2C0 defaults,
   SCL=Pin(GP9), SDA=Pin(GP8), freq=400000
10 print("I2C Address      : "+hex(i2c.scan()[0]).upper()) # Display device address
11 print("I2C Configuration: "+str(i2c))                   # Display I2C config
12
```

```python
13
14 oled = SSD1306_I2C(WIDTH, HEIGHT, i2c)              # Init oled display
15
16 # Raspberry Pi logo as 32x32 bytearray
17 buffer = bytearray(b"\x00\x00\x00\x00\x00\x00\x00\x00\x00\x00\x00\x00\x00|?\x00\x01\x86
   @\x80\x01\x01\x80\x80\x01\x11\x88\x80\x01\x05\xa0\x80\x00\x83\xc1\x00\x00C\xe3\x00\x00
   ~\xfc\x00\x00L'\x00\x00\x9c\x11\x00\x00\xbf\xfd\x00\x00\xe1\x87\x00\x01\xc1\x83\x80\x02A
   \x82@\x02A\x82@\x02\xc1\xc2@\x02\xf6>\xc0\x01\xfc
   =\x80\x01\x18\x18\x80\x01\x88\x10\x80\x00\x8c!\x00\x00\x87\xf1\x00\x00\x7f\xf6\x00\x00
   8\x1c\x00\x00\x0c \x00\x00\x03\xc0\x00\x00\x00\x00\x00\x00\x00\x00\x00\x00\x00\x00")
18
19 # Load the raspberry pi logo into the framebuffer (the image is 32x32)
20 fb = framebuf.FrameBuffer(buffer, 32, 32, framebuf.MONO_HLSB)
21
22 # Clear the oled display in case it has junk on it.
23 oled.fill(0)
24
25 # Blit the image from the framebuffer to the oled display
26 oled.blit(fb, 96, 0)
27
28 # Add some text
29 oled.text("Raspberry Pi",5,5)
30 oled.text("Pico",5,15)
31
32 # Finally update the oled display so the image & text is displayed
33 oled.show()
```

i2c_1306oled_with_freq.py

The example code, explicitly sets a frequency.

Pico MicroPython Examples: https://github.com/raspberrypi/pico-micropython-examples/blob/master/i2c/1306oled/i2c_1306oled_with_freq.py

```python
1  # Display Image & text on I2C driven ssd1306 OLED display
2  from machine import Pin, I2C
3  from ssd1306 import SSD1306_I2C
4  import framebuf
5
6  WIDTH  = 128                                         # oled display width
7  HEIGHT = 32                                          # oled display height
8
9  i2c = I2C(0, scl=Pin(9), sda=Pin(8), freq=200000)    # Init I2C using pins GP8 & GP9
   (default I2C0 pins)
10 print("I2C Address      : "+hex(i2c.scan()[0]).upper()) # Display device address
11 print("I2C Configuration: "+str(i2c))                # Display I2C config
12
13
14 oled = SSD1306_I2C(WIDTH, HEIGHT, i2c)              # Init oled display
15
16 # Raspberry Pi logo as 32x32 bytearray
17 buffer = bytearray(b"\x00\x00\x00\x00\x00\x00\x00\x00\x00\x00\x00\x00\x00|?\x00\x01\x86
   @\x80\x01\x01\x80\x80\x01\x11\x88\x80\x01\x05\xa0\x80\x00\x83\xc1\x00\x00C\xe3\x00\x00
   ~\xfc\x00\x00L'\x00\x00\x9c\x11\x00\x00\xbf\xfd\x00\x00\xe1\x87\x00\x01\xc1\x83\x80\x02A
   \x82@\x02A\x82@\x02\xc1\xc2@\x02\xf6>\xc0\x01\xfc
   =\x80\x01\x18\x18\x80\x01\x88\x10\x80\x00\x8c!\x00\x00\x87\xf1\x00\x00\x7f\xf6\x00\x00
   8\x1c\x00\x00\x0c \x00\x00\x03\xc0\x00\x00\x00\x00\x00\x00\x00\x00\x00\x00\x00\x00")
18
19 # Load the raspberry pi logo into the framebuffer (the image is 32x32)
20 fb = framebuf.FrameBuffer(buffer, 32, 32, framebuf.MONO_HLSB)
21
22 # Clear the oled display in case it has junk on it.
23 oled.fill(0)
```

```
24
25  # Blit the image from the framebuffer to the oled display
26  oled.blit(fb, 96, 0)
27
28  # Add some text
29  oled.text("Raspberry Pi",5,5)
30  oled.text("Pico",5,15)
31
32  # Finally update the oled display so the image & text is displayed
33  oled.show()
```

Bill of Materials

Table 4. A list of materials required for the example

Item	Quantity	Details
Breadboard	1	generic part
Raspberry Pi Pico	1	https://www.raspberrypi.com/products/raspberry-pi-pico/
Monochrome 128x32 I2C OLED Display	1	https://www.adafruit.com/product/931

Using a SH1106-based OLED graphics display

Display an image and text on I2C driven SH1106-based OLED graphics display such as the Pimoroni Breakout Garden 1.12" Mono OLED https://shop.pimoroni.com/products/1-12-oled-breakout?variant=29421050757203 .

Wiring information

See Figure 7 for wiring instructions.

Figure 7. Wiring the OLED to Pico using I2C

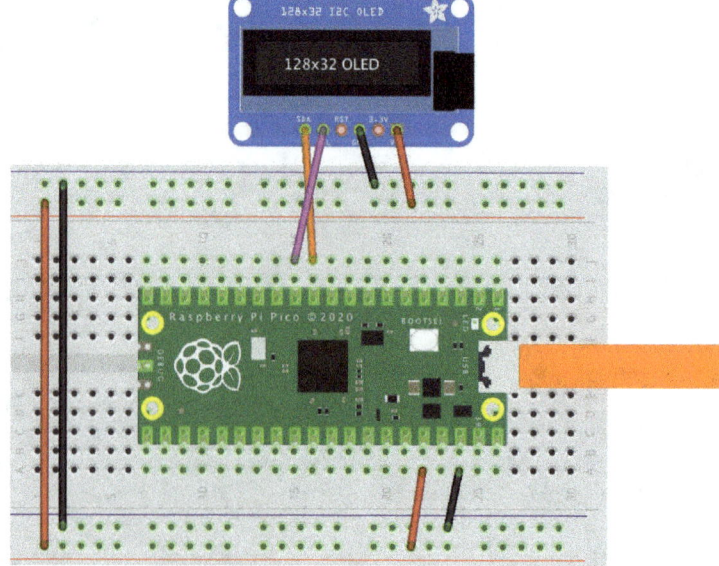

List of Files

A list of files with descriptions of their function;

i2c_1106oled_using_defaults.py

The example code.

Pico MicroPython Examples: https://github.com/raspberrypi/pico-micropython-examples/blob/master/i2c/1106oled/i2c_1106oled_using_defaults.py

```python
 1  # Display Image & text on I2C driven SH1106 OLED display
 2  from machine import I2C, ADC
 3  from sh1106 import SH1106_I2C
 4  import framebuf
 5
 6
 7  WIDTH  = 128                                            # oled display width
 8  HEIGHT = 128                                            # oled display height
 9
10  i2c = I2C(0)                                            # Init I2C using I2C0 defaults,
    SCL=Pin(GP9), SDA=Pin(GP8), freq=400000
11  print("I2C Address      : "+hex(i2c.scan()[0]).upper()) # Display device address
12  print("I2C Configuration: "+str(i2c))                   # Display I2C config
13
14
15  oled = SH1106_I2C(WIDTH, HEIGHT, i2c)                   # Init oled display
16
17  # Raspberry Pi logo as 32x32 bytearray
18  buffer = bytearray(b"\x00\x00\x00\x00\x00\x00\x00\x00\x00\x00\x00\x00\x00|?\x00\x01\x86
    @\x80\x01\x01\x80\x80\x01\x11\x88\x80\x01\x05\xa0\x80\x00\x83\xc1\x00\x00C\xe3\x00\x00
    ~\xfc\x00\x00L'\x00\x00\x9c\x11\x00\x00\xbf\xfd\x00\x00\xe1\x87\x00\x01\xc1\x83\x80\x02A
    \x82@\x02A\x82@\x02\xc1\xc2@\x02\xf6>\xc0\x01\xfc
    =\x80\x01\x18\x18\x80\x01\x88\x10\x80\x00\x8c!\x00\x00\x87\xf1\x00\x00\x7f\xf6\x00\x00
    8\x1c\x00\x00\x0c \x00\x00\x03\xc0\x00\x00\x00\x00\x00\x00\x00\x00\x00\x00\x00\x00")
19
20  # Load the raspberry pi logo into the framebuffer (the image is 32x32)
21  fb = framebuf.FrameBuffer(buffer, 32, 32, framebuf.MONO_HLSB)
22
23  # Clear the oled display in case it has junk on it.
24  oled.fill(0)
25
26  # Blit the image from the framebuffer to the oled display
27  oled.blit(fb, 96, 0)
28
29  # Add some text
30  oled.text("Raspberry Pi",5,5)
31  oled.text("Pico",5,15)
32
33  # Finally update the oled display so the image & text is displayed
34  oled.show()
```

i2c_1106oled_with_freq.py

The example code, explicitly sets a frequency.

Pico MicroPython Examples: https://github.com/raspberrypi/pico-micropython-examples/blob/master/i2c/1106oled/i2c_1106oled_with_freq.py

```python
1  # Display Image & text on I2C driven ssd1306 OLED display
2  from machine import Pin, I2C
3  from sh1106 import SH1106_I2C
4  import framebuf
5
6  WIDTH  = 128                                            # oled display width
```

```python
7  HEIGHT = 32                                          # oled display height
8
9  i2c = I2C(0, scl=Pin(9), sda=Pin(8), freq=200000)    # Init I2C using pins GP8 & GP9
   (default I2C0 pins)
10 print("I2C Address      : "+hex(i2c.scan()[0]).upper()) # Display device address
11 print("I2C Configuration: "+str(i2c))                # Display I2C config
12
13
14 oled = SH1106_I2C(WIDTH, HEIGHT, i2c)                # Init oled display
15
16 # Raspberry Pi logo as 32x32 bytearray
17 buffer = bytearray(b"\x00\x00\x00\x00\x00\x00\x00\x00\x00\x00\x00\x00\x00|?\x00\x01\x86
   @\x80\x01\x01\x80\x80\x01\x11\x88\x80\x01\x05\xa0\x80\x00\x83\xc1\x00\x00C\xe3\x00\x00
   ~\xfc\x00\x00L'\x00\x00\x9c\x11\x00\x00\xbf\xfd\x00\x00\xe1\x87\x00\x01\xc1\x83\x80\x02A
   \x82@\x02A\x82@\x02\xc1\xc2@\x02\xf6>\xc0\x01\xfc
   =\x80\x01\x18\x18\x80\x01\x88\x10\x80\x00\x8c!\x00\x00\x87\xf1\x00\x00\x7f\xf6\x00\x00
   8\x1c\x00\x00\x0c \x00\x00\x03\xc0\x00\x00\x00\x00\x00\x00\x00\x00\x00\x00\x00\x00")
18
19 # Load the raspberry pi logo into the framebuffer (the image is 32x32)
20 fb = framebuf.FrameBuffer(buffer, 32, 32, framebuf.MONO_HLSB)
21
22 # Clear the oled display in case it has junk on it.
23 oled.fill(0)
24
25 # Blit the image from the framebuffer to the oled display
26 oled.blit(fb, 96, 0)
27
28 # Add some text
29 oled.text("Raspberry Pi",5,5)
30 oled.text("Pico",5,15)
31
32 # Finally update the oled display so the image & text is displayed
33 oled.show()
```

sh1106.py

SH1106 Driver Obtained from https://github.com/robert-hh/SH1106

Pico MicroPython Examples: https://github.com/raspberrypi/pico-micropython-examples/blob/master/i2c/1106oled/sh1106.py

```
1  #
2  # MicroPython SH1106 OLED driver, I2C and SPI interfaces
3  #
4  # The MIT License (MIT)
5  #
6  # Copyright (c) 2016 Radomir Dopieralski (@deshipu),
7  #               2017 Robert Hammelrath (@robert-hh)
8  #
9  # Permission is hereby granted, free of charge, to any person obtaining a copy
10 # of this software and associated documentation files (the "Software"), to deal
11 # in the Software without restriction, including without limitation the rights
12 # to use, copy, modify, merge, publish, distribute, sublicense, and/or sell
13 # copies of the Software, and to permit persons to whom the Software is
14 # furnished to do so, subject to the following conditions:
15 #
16 # The above copyright notice and this permission notice shall be included in
17 # all copies or substantial portions of the Software.
18 #
19 # THE SOFTWARE IS PROVIDED "AS IS", WITHOUT WARRANTY OF ANY KIND, EXPRESS OR
20 # IMPLIED, INCLUDING BUT NOT LIMITED TO THE WARRANTIES OF MERCHANTABILITY,
21 # FITNESS FOR A PARTICULAR PURPOSE AND NONINFRINGEMENT. IN NO EVENT SHALL THE
22 # AUTHORS OR COPYRIGHT HOLDERS BE LIABLE FOR ANY CLAIM, DAMAGES OR OTHER
```

```python
23 # LIABILITY, WHETHER IN AN ACTION OF CONTRACT, TORT OR OTHERWISE, ARISING FROM,
24 # OUT OF OR IN CONNECTION WITH THE SOFTWARE OR THE USE OR OTHER DEALINGS IN
25 # THE SOFTWARE.
26 #
27 # Sample code sections
28 # ------------ SPI ------------------
29 # Pin Map SPI
30 #   - 3V3        - Vcc
31 #   - GND        - Gnd
32 #   - GPIO 11    - DIN / MOSI fixed
33 #   - GPIO 10    - CLK / Sck fixed
34 #   - GPIO 4     - CS (optional, if the only connected device, connect to GND)
35 #   - GPIO 5     - D/C
36 #   - GPIO 2     - Res
37 #
38 # for CS, D/C and Res other ports may be chosen.
39 #
40 # from machine import Pin, SPI
41 # import sh1106
42
43 # spi = SPI(1, baudrate=1000000)
44 # display = sh1106.SH1106_SPI(128, 64, spi, Pin(5), Pin(2), Pin(4))
45 # display.sleep(False)
46 # display.fill(0)
47 # display.text('Testing 1', 0, 0, 1)
48 # display.show()
49 #
50 # --------------- I2C ------------------
51 #
52 # Pin Map I2C
53 #   - 3V3        - Vcc
54 #   - GND        - Gnd
55 #   - GPIO 5     - CLK / SCL
56 #   - GPIO 4     - DIN / SDA
57 #   - GPIO 2     - Res
58 #   - GND        - CS
59 #   - GND        - D/C
60 #
61 #
62 # from machine import Pin, I2C
63 # import sh1106
64 #
65 # i2c = I2C(0, scl=Pin(5), sda=Pin(4), freq=400000)
66 # display = sh1106.SH1106_I2C(128, 64, i2c, Pin(2), 0x3c)
67 # display.sleep(False)
68 # display.fill(0)
69 # display.text('Testing 1', 0, 0, 1)
70 # display.show()
71
72 from micropython import const
73 import utime as time
74 import framebuf
75
76
77 # a few register definitions
78 _SET_CONTRAST        = const(0x81)
79 _SET_NORM_INV        = const(0xa6)
80 _SET_DISP            = const(0xae)
81 _SET_SCAN_DIR        = const(0xc0)
82 _SET_SEG_REMAP       = const(0xa0)
83 _LOW_COLUMN_ADDRESS  = const(0x00)
84 _HIGH_COLUMN_ADDRESS = const(0x10)
85 _SET_PAGE_ADDRESS    = const(0xB0)
```

```python
86
87
88 class SH1106:
89     def __init__(self, width, height, external_vcc):
90         self.width = width
91         self.height = height
92         self.external_vcc = external_vcc
93         self.pages = self.height // 8
94         self.buffer = bytearray(self.pages * self.width)
95         fb = framebuf.FrameBuffer(self.buffer, self.width, self.height,
96                                   framebuf.MVLSB)
97         self.framebuf = fb
98         # set shortcuts for the methods of framebuf
99         self.fill = fb.fill
100        self.fill_rect = fb.fill_rect
101        self.hline = fb.hline
102        self.vline = fb.vline
103        self.line = fb.line
104        self.rect = fb.rect
105        self.pixel = fb.pixel
106        self.scroll = fb.scroll
107        self.text = fb.text
108        self.blit = fb.blit
109
110        self.init_display()
111
112    def init_display(self):
113        self.reset()
114        self.fill(0)
115        self.poweron()
116        self.show()
117
118    def poweroff(self):
119        self.write_cmd(_SET_DISP | 0x00)
120
121    def poweron(self):
122        self.write_cmd(_SET_DISP | 0x01)
123
124    def rotate(self, flag, update=True):
125        if flag:
126            self.write_cmd(_SET_SEG_REMAP | 0x01)  # mirror display vertically
127            self.write_cmd(_SET_SCAN_DIR | 0x08)   # mirror display hor.
128        else:
129            self.write_cmd(_SET_SEG_REMAP | 0x00)
130            self.write_cmd(_SET_SCAN_DIR | 0x00)
131        if update:
132            self.show()
133
134    def sleep(self, value):
135        self.write_cmd(_SET_DISP | (not value))
136
137    def contrast(self, contrast):
138        self.write_cmd(_SET_CONTRAST)
139        self.write_cmd(contrast)
140
141    def invert(self, invert):
142        self.write_cmd(_SET_NORM_INV | (invert & 1))
143
144    def show(self):
145        for page in range(self.height // 8):
146            self.write_cmd(_SET_PAGE_ADDRESS | page)
147            self.write_cmd(_LOW_COLUMN_ADDRESS | 2)
148            self.write_cmd(_HIGH_COLUMN_ADDRESS | 0)
```

```python
149            self.write_data(self.buffer[
150                self.width * page:self.width * page + self.width
151            ])
152
153    def reset(self, res):
154        if res is not None:
155            res(1)
156            time.sleep_ms(1)
157            res(0)
158            time.sleep_ms(20)
159            res(1)
160            time.sleep_ms(20)
161
162
163 class SH1106_I2C(SH1106):
164    def __init__(self, width, height, i2c, res=None, addr=0x3c,
165                 external_vcc=False):
166        self.i2c = i2c
167        self.addr = addr
168        self.res = res
169        self.temp = bytearray(2)
170        if res is not None:
171            res.init(res.OUT, value=1)
172        super().__init__(width, height, external_vcc)
173
174    def write_cmd(self, cmd):
175        self.temp[0] = 0x80  # Co=1, D/C#=0
176        self.temp[1] = cmd
177        self.i2c.writeto(self.addr, self.temp)
178
179    def write_data(self, buf):
180        self.i2c.writeto(self.addr, b'\x40'+buf)
181
182    def reset(self):
183        super().reset(self.res)
184
185
186 class SH1106_SPI(SH1106):
187    def __init__(self, width, height, spi, dc, res=None, cs=None,
188                 external_vcc=False):
189        self.rate = 10 * 1000 * 1000
190        dc.init(dc.OUT, value=0)
191        if res is not None:
192            res.init(res.OUT, value=0)
193        if cs is not None:
194            cs.init(cs.OUT, value=1)
195        self.spi = spi
196        self.dc = dc
197        self.res = res
198        self.cs = cs
199        super().__init__(width, height, external_vcc)
200
201    def write_cmd(self, cmd):
202        self.spi.init(baudrate=self.rate, polarity=0, phase=0)
203        if self.cs is not None:
204            self.cs(1)
205            self.dc(0)
206            self.cs(0)
207            self.spi.write(bytearray([cmd]))
208            self.cs(1)
209        else:
210            self.dc(0)
211            self.spi.write(bytearray([cmd]))
```

```
212
213     def write_data(self, buf):
214         self.spi.init(baudrate=self.rate, polarity=0, phase=0)
215         if self.cs is not None:
216             self.cs(1)
217             self.dc(1)
218             self.cs(0)
219             self.spi.write(buf)
220             self.cs(1)
221         else:
222             self.dc(1)
223             self.spi.write(buf)
224
225     def reset(self):
226         super().reset(self.res)
```

Bill of Materials

Table 5. A list of materials required for the example

Item	Quantity	Details
Breadboard	1	generic part
Raspberry Pi Pico	1	https://www.raspberrypi.com/products/raspberry-pi-pico/
Monochrome 128x128 I2C OLED Display	1	https://shop.pimoroni.com/products/1-12-oled-breakout?variant=29421050757203

Using PIO to drive a set of NeoPixel Ring (WS2812 LEDs)

Combination of the PIO WS2812 demo with the Adafruit 'essential' NeoPixel example code to show off color fills, chases and of course a rainbow swirl on a 16-LED ring.

Wiring information

See Figure 8 for wiring instructions.

Figure 8. Wiring the 16-LED NeoPixel Ring to Pico

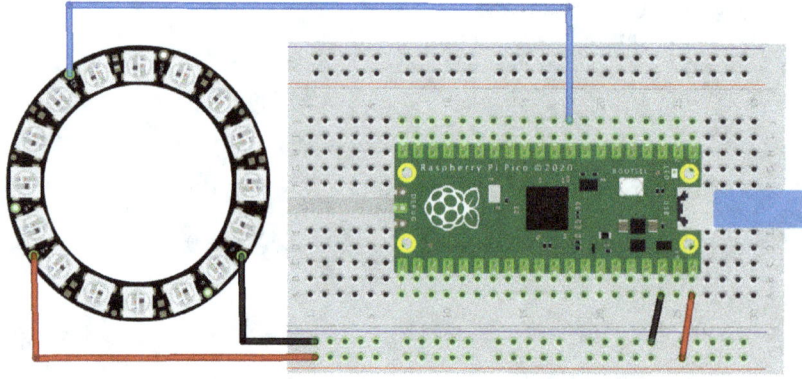

List of Files

A list of files with descriptions of their function;

neopixel_ring.py

The example code.

Pico MicroPython Examples: https://github.com/raspberrypi/pico-micropython-examples/blob/master/pio/neopixel_ring/neopixel_ring.py

```python
 1  # Example using PIO to drive a set of WS2812 LEDs.
 2
 3  import array, time
 4  from machine import Pin
 5  import rp2
 6
 7  # Configure the number of WS2812 LEDs.
 8  NUM_LEDS = 16
 9  PIN_NUM = 6
10  brightness = 0.2
11
12  @rp2.asm_pio(sideset_init=rp2.PIO.OUT_LOW, out_shiftdir=rp2.PIO.SHIFT_LEFT, autopull=True, pull_thresh=24)
13  def ws2812():
14      T1 = 2
15      T2 = 5
16      T3 = 3
17      wrap_target()
18      label("bitloop")
19      out(x, 1)               .side(0)    [T3 - 1]
20      jmp(not_x, "do_zero")   .side(1)    [T1 - 1]
21      jmp("bitloop")          .side(1)    [T2 - 1]
22      label("do_zero")
23      nop()                   .side(0)    [T2 - 1]
24      wrap()
25
26
27  # Create the StateMachine with the ws2812 program, outputting on pin
28  sm = rp2.StateMachine(0, ws2812, freq=8_000_000, sideset_base=Pin(PIN_NUM))
29
30  # Start the StateMachine, it will wait for data on its FIFO.
31  sm.active(1)
32
33  # Display a pattern on the LEDs via an array of LED RGB values.
34  ar = array.array("I", [0 for _ in range(NUM_LEDS)])
35
36  ########################################################################
37  def pixels_show():
38      dimmer_ar = array.array("I", [0 for _ in range(NUM_LEDS)])
39      for i,c in enumerate(ar):
40          r = int(((c >> 8) & 0xFF) * brightness)
41          g = int(((c >> 16) & 0xFF) * brightness)
42          b = int((c & 0xFF) * brightness)
43          dimmer_ar[i] = (g<<16) + (r<<8) + b
44      sm.put(dimmer_ar, 8)
45      time.sleep_ms(10)
46
47  def pixels_set(i, color):
48      ar[i] = (color[1]<<16) + (color[0]<<8) + color[2]
49
50  def pixels_fill(color):
51      for i in range(len(ar)):
52          pixels_set(i, color)
```

```
 53
 54 def color_chase(color, wait):
 55     for i in range(NUM_LEDS):
 56         pixels_set(i, color)
 57         time.sleep(wait)
 58         pixels_show()
 59     time.sleep(0.2)
 60
 61 def wheel(pos):
 62     # Input a value 0 to 255 to get a color value.
 63     # The colours are a transition r - g - b - back to r.
 64     if pos < 0 or pos > 255:
 65         return (0, 0, 0)
 66     if pos < 85:
 67         return (255 - pos * 3, pos * 3, 0)
 68     if pos < 170:
 69         pos -= 85
 70         return (0, 255 - pos * 3, pos * 3)
 71     pos -= 170
 72     return (pos * 3, 0, 255 - pos * 3)
 73
 74
 75 def rainbow_cycle(wait):
 76     for j in range(255):
 77         for i in range(NUM_LEDS):
 78             rc_index = (i * 256 // NUM_LEDS) + j
 79             pixels_set(i, wheel(rc_index & 255))
 80         pixels_show()
 81         time.sleep(wait)
 82
 83 BLACK = (0, 0, 0)
 84 RED = (255, 0, 0)
 85 YELLOW = (255, 150, 0)
 86 GREEN = (0, 255, 0)
 87 CYAN = (0, 255, 255)
 88 BLUE = (0, 0, 255)
 89 PURPLE = (180, 0, 255)
 90 WHITE = (255, 255, 255)
 91 COLORS = (BLACK, RED, YELLOW, GREEN, CYAN, BLUE, PURPLE, WHITE)
 92
 93 print("fills")
 94 for color in COLORS:
 95     pixels_fill(color)
 96     pixels_show()
 97     time.sleep(0.2)
 98
 99 print("chases")
100 for color in COLORS:
101     color_chase(color, 0.01)
102
103 print("rainbow")
104 rainbow_cycle(0)
```

Bill of Materials

Table 6. A list of materials required for the example

Item	Quantity	Details
Breadboard	1	generic part

Raspberry Pi Pico	1	https://www.raspberrypi.com/products/raspberry-pi-pico/
NeoPixel Ring	1	https://www.adafruit.com/product/1463

Using UART on the Raspberry Pi Pico

Send data from the UART1 port to the UART0 port. Other things to try;

```
uart0 = UART(0)
```

which will open a UART connection at the default baudrate of 115200, and

```
uart0.readline()
```

which will read until the CR (\r) and NL (\n) characters, then return the line.

Wiring information

See Figure 9 for wiring instructions.

Figure 9. Wiring two of the Pico's ports together. Be sure to wire UART0 TX to UART1 RX and UART0 RX to UART1 TX.

List of Files

A list of files with descriptions of their function;

uart.py
    The example code.

Pico MicroPython Examples: *https://github.com/raspberrypi/pico-micropython-examples/blob/master/uart/loopback/uart.py*

```python
from machine import UART, Pin
import time

uart1 = UART(1, baudrate=9600, tx=Pin(8), rx=Pin(9))

uart0 = UART(0, baudrate=9600, tx=Pin(0), rx=Pin(1))

txData = b'hello world\n\r'
uart1.write(txData)
time.sleep(0.1)
rxData = bytes()
while uart0.any() > 0:
    rxData += uart0.read(1)

print(rxData.decode('utf-8'))
```

Bill of Materials

Table 7. A list of materials required for the example

Item	Quantity	Details
Breadboard	1	generic part
Raspberry Pi Pico	1	https://www.raspberrypi.com/products/raspberry-pi-pico/

Appendix B: Documentation Release History

Table 8. Documentation release history

Release	Date	Description
1.0	21 Jan 2021	- Initial release
1.1	26 Jan 2021	- Minor corrections - Extra information about using DMA with ADC - Clarified M0+ and SIO CPUID registers - Added more discussion of Timers - Update Windows and macOS build instructions - Renamed books and optimised size of output PDFs
1.2	01 Feb 2021	- Minor corrections - Small improvements to PIO documentation - Added missing TIMER2 and TIMER3 registers to DMA - Explained how to get MicroPython REPL on UART - To accompany the V1.0.1 release of the C SDK
1.3	23 Feb 2021	- Minor corrections - Changed font - Additional documentation on sink/source limits for RP2040 - Major improvements to SWD documentation - Updated MicroPython build instructions - MicroPython UART example code - Updated Thonny instructions - Updated Project Generator instructions - Added a FAQ document - Added errata E7, E8 and E9
1.3.1	05 Mar 2021	- Minor corrections - To accompany the V1.1.0 release of the C SDK - Improved MicroPython UART example - Improved Pinout diagram
1.4	07 Apr 2021	- Minor corrections - Added errata E10 - Note about how to update the C SDK from Github - To accompany the V1.1.2 release of the C SDK

Release	Date	Description
1.4.1	13 Apr 2021	• Minor corrections • Clarified that all source code in the documentation is under the 3-Clause BSD license.
1.5	07 Jun 2021	• Minor updates and corrections • Updated FAQ • Added SDK release history • To accompany the V1.2.0 release of the C SDK
1.6	23 Jun 2021	• Minor updates and corrections • ADC information updated • Added errata E11
1.6.1	30 Sep 2021	• Minor updates and corrections • Information about B2 release • Updated errata for B2 release
1.7	03 Nov 2021	• Minor updates and corrections • Fixed some register access types and descriptions • Added core 1 launch sequence info • Described SDK "panic" handling • Updated picotool documentation • Additional examples added to **Appendix A: App Notes** appendix in the Raspberry Pi Pico C/C++ SDK book • To accompany the V1.3.0 release of the C SDK
1.7.1	04 Nov 2021	• Minor updates and corrections • Better documentation of USB double buffering • Picoprobe branch changes • Updated links to documentation
1.8	17 Jun 2022	• Minor updates and corrections • Updated setup instructions for Windows in Getting started with Raspberry Pi Pico • Additional explanation of SDK configuration • RP2040 now qualified to -40°C, minimum operating temperature changed from -20°C to -40°C • Increased PLL min VCO from 400MHz to 750MHz for improved stability across operating conditions • Added reflow-soldering temperature profile • Added errata E12, E13 and E14 • To accompany the V1.3.1 release of the C SDK

Release	Date	Description
1.9	30 Jun 2022	- Minor updates and corrections - Update to VGA board hardware description for launch of Raspberry Pi Pico W - To accompany the V1.4.0 release of the C SDK
Pico and Pico W databooks combined into a unified release history		
2.0	01 Dec 2022	- Minor updates and corrections - Added RP2040 availability information - Added RP2040 storage conditions and thermal characteristics - Replace SDK library documentation with links to the online version - Updated Picoprobe build and usage instructions

The latest release can be found at https://datasheets.raspberrypi.com/pico/raspberry-pi-pico-python-sdk.pdf.

Raspberry Pi is a trademark of Raspberry Pi Ltd

This page was intentionally left blank.

This page was intentionally left blank.